PRAWN FARMING FOR BEGINNERS

A Complete Step-By-Step Guide To Sustainable Techniques, Pond Management, Feeding, And Maximizing Profits

Holden bodhi

Contents

CHAPTER ONE 10
  Overview Of Prawn Farming 10
    An Overview Of Farming Prawns 10
    Advantages And Difficulties 11
    Difficulties In Growing Prawns 12
    Important Words 13

CHAPTER TWO 17
  Knowing About Prawn Species 17
    Typical Prawn Breeds For Agriculture 17
    Qualities And Prerequisites 20
    Selecting The Proper Species 22

CHAPTER THREE 26
  Organizing Your Farm For Prawns 26
    Choosing A Farm Location 26
    Pond Design And Construction 28
    Management Of Water Quality 30

CHAPTER FOUR 34
  Basics Of Prawn Hatchery 34
    Equipment And Setup For Hatcheries 34
    The Selection And Management Of Broodstock 37
    Management Of Stocking And Nurseries 40

CHAPTER FIVE 43
  Nutrition And Feeding In Prawn Farming 43
    Prawn Feed Types 43

. Feeding Methods And Schedules 45

Dietary Needs 48

CHAPTER SIX 51

Nutrition And Feeding 51

Prawn Feed Types 51

Feeding Methods And Schedules 53

Dietary Needs 55

CHAPTER SEVEN 59

Water Resources Management 59

Parameters Of Water Quality 59

Systems For Filtration And Aeration 62

Frequent Inspection And Upkeep 65

CHAPTER EIGHT 68

Water Resources Management 68

Overview of Water Management in Prawn Production .. 68

Systems For Filtration And Aeration 70

Frequent Inspection And Upkeep 72

CHAPTER NINE 75

Water Resources Management 75

Parameters Of Water Quality 75

Systems For Filtration And Aeration 78

Frequent Inspection And Upkeep 81

CHAPTER TEN 84

Legal And Environmental Aspects 84

Sustainability And Its Effect On The Environment...........84
Eco-Friendly Meal Techniques..87
Standards For Food Safety And Quality 89
CHAPTER ELEVEN...94
Increasing The Size Of Your Prawn Farm 94
Planning And Growth Strategies 94
Financial Management And Investment.........................97
..Investigating Novel Markets And Prospects 100
Conclusion 103
THE END 108

Copyright © 2024 by Holden bodhi

All rights reserved.

No part of this publication may be reproduced, distributed, or transmitted in any form or by any means, including photocopying, recording, or other electronic or mechanical methods, without the prior written permission of the publisher, except in the case of brief quotations embodied in critical reviews and certain other non

commercial uses permitted by copyright law.

DISCLAIMER

The information provided in this book, is intended for educational and informational purposes only. The content is based on research, personal experiences, and general knowledge about farming. It is not intended to substitute professional advice or expert consultation. Readers are encouraged to seek professional guidance when implementing any practices or techniques discussed in this book.

The author and publisher make no representations or warranties of any kind regarding the accuracy, applicability, or completeness of the contents of this book. Any reliance you place on such information is strictly at your own risk. The author and publisher shall not be held liable for any damages, losses, or injuries resulting from the use of the information provided.

Additionally, the author does not endorse, recommend, or affiliate with any individual, product, service, website, organization, or brand mentioned or referenced in this book. Any such references are solely for informational purposes, and no warranty or guarantee is implied. The inclusion of these references does not imply any endorsement or partnership by the author.

By reading this book, you acknowledge and accept that the author and publisher are not responsible for any consequences arising from your use of the information provided.

CHAPTER ONE

Overview Of Prawn Farming

An Overview Of Farming Prawns

Shrimp farming, or "prawn farming," is an important subset of aquaculture that focuses on the commercial prawn growing industry. Due to the rising demand for fish across the world in recent decades, this process has undergone major change. In contrast to conventional fishing, which has the potential to deplete natural populations, prawn farming provides a sustainable means of producing this important food.

The process of raising prawns in captivity, commonly in ponds or tanks, entails providing the best circumstances possible for their growth. These settings are intended to closely resemble natural habitats while giving farmers the ability to keep an eye on and regulate a variety of parameters, including feeding schedules, disease management, and water quality. Prawn farms may be small-scale family businesses or major commercial operations, depending on the resources and demands of the market.

Advantages And Difficulties
Advantages of Farming Prawns

1. Possibilities for Economic Gain: Prawn farming has the potential to provide substantial financial gains, particularly in coastal regions where conventional fishing practices may be waning. It boosts regional economies by generating employment in cultivation, processing, and distribution.

2. Sustainable Seafood Supply: The strain on wild populations is lessened by raising prawns in regulated habitats. In addition to ensuring a consistent supply of seafood, this helps maintain natural habitats.

3. Modern prawn farming makes use of cutting-edge technology for feeding, water management, and health monitoring. Better-quality prawns and increased production efficiency result from this.

4. Market Versatility: Prawns are highly prized in international marketplaces, and demand for them never fluctuates due to how well they cook. Depending on their size and resources, farmers might focus on regional or global markets.

Difficulties In Growing Prawns

1. Environmental Concerns: If prawn farms are not correctly maintained, they may have an adverse effect on the surrounding ecosystems. Sustainable practices are crucial because they may prevent problems like disease outbreaks, habitat damage, and water contamination.

2. Management of Diseases: Prawns are vulnerable to a wide range of illnesses and

pests. Efficient disease control is essential to avert epidemics that might destroy agricultural communities and cause economic damage.

3. Economic Fluctuations: A number of variables, including global trade laws, consumer demand, and manufacturing expenses, may affect the prawn market's volatility. To continue to be profitable, farmers must manage these swings.

4. Resource-intensive: Raising prawns requires a lot of resources, such as energy, feed, and water of the highest caliber. One of the main challenges is making sure that these resources are used efficiently while minimizing waste.

Important Words

Pond System: A popular technique for raising prawns in which the prawns are raised in ponds that have been manmade. These ponds come in a range of sizes and levels of sophistication, from basic earthen ponds to sophisticated walled ponds with automated water control systems.

Biosecurity refers to the procedures and policies put in place to stop pests and illnesses from entering and spreading across prawn farms. This includes following quarantine procedures, doing routine health examinations, and keeping surroundings and equipment clean.

The term "stocking density" describes how many prawns are positioned in a certain area within a farming system. Controlling the density of the stocking is essential to guaranteeing maximum development and reducing stress levels in the prawns.

Monitoring and regulating water quality involves keeping an eye on factors including temperature, salinity, pH, and oxygen content. Prawn growth and health depend on excellent water quality being maintained.

Harvesting is the process of removing fully grown prawns from the agricultural system. To guarantee that prawns are handled carefully and effectively, maintaining their quality for the market, proper harvesting practices are crucial.

Feeding Regime: The strategy and techniques used to provide prawns nourishment. This covers the kind, frequency, and quantity of feeding that are appropriate for the prawns' particular development stage and requirements.

Recirculating Aquaculture System (RAS): An advanced farming technique that minimizes its effect on the environment and lessens the demand for new water by recycling water inside the farm. To preserve the prawns' health and the water's purity, RAS systems need to be carefully managed.

Grow-out Period: The amount of time from farm stocking prawns to harvesting them. Depending on the prawn species, farming circumstances, and target market size, this time frame may change.

Disease management refers to the methods and techniques used in prawn farms to track, stop, and manage infections. This includes immunization, routine physical examinations, and the use of therapeutic interventions.

Sustainable practices are strategies and tactics used to ensure long-term production while reducing the negative effects of prawn farming on the environment. This entails using sustainable farming practices, cutting waste, and utilizing eco-friendly feed.

Each of these components is essential to the smooth running of a prawn farming enterprise. Gaining an understanding of these ideas is essential for anybody wishing to work in the field and guarantees that agricultural methods benefit the market and the environment.

CHAPTER TWO

Knowing About Prawn Species

Typical Prawn Breeds For Agriculture

Raising prawns for commercial use in controlled conditions is known as aquaculture or prawn farming. The proper species must be chosen for a prawn farm to be successful since each has distinct traits and needs different conditions. Gaining an understanding of these species is essential to maximizing profitability, growth, and production.

1. Penaeus vannamei, or white-leg prawn

One of the most widely farmed species in aquaculture worldwide is the Whiteleg shrimp, often referred to as Pacific white shrimp. This species, which is native to the Pacific coasts of Central and South America, is renowned for growing quickly and adapting to a variety of environments. It is appropriate for brackish as well as marine habitats because of its broad tolerance of salt levels.

2. Penaeus monodon, or black tiger shrimp

In terms of prawn farming, another important species is the Black Tiger prawn. It comes from the Indo-Pacific area and is easily identified by its big size and characteristic black stripes. Compared to Whiteleg prawns, this species needs more salinity and consistent water conditions. Due to its size and flavor, it is highly prized in the market; yet, it could be more susceptible to changes in the surrounding environment.

3. The macrobrachrachium rosenbergii, or giant river prawn

The Giant River Prawn is indigenous to freshwater and brackish water systems in Southeast Asia, in contrast to marine species. It is a favorite in freshwater aquaculture, despite its reputation for aggression and size.

Compared to marine species, this species has a more complicated breeding cycle and needs a certain range of pH and water temperature.

4. Penaeus japonicus, or Kuruma Prawn

China and Japan's coastal seas are home to the Kuruma prawn. Due to its particular ecological needs, it is often farmed in highly specialized systems and is well-known for its high market value. This species needs strict water quality control since it does well in warmer climates.

5. Penguin (Penaeus merguiensis) prawn

The Indo-Pacific area is home to the banana prawn, which is distinguished by its unusual golden hue. Though smaller in size than Whiteleg and Black Tiger prawns, this species is prized for its flavor and texture. It is not as tolerant to low salinity levels as the Whiteleg prawn, but it does need comparable environmental conditions.

Qualities And Prerequisites
1. Salinity and Temperature

The preferred salinity and temperature of different prawn species differ. For example, the Whiteleg prawn can survive in a broad range of salinities (0.5–30 ppt) and temperatures (20–30 °C). The Black Tiger prawn, on the other hand, likes steady temperatures of 28 to 30°C and higher salinities. To create the ideal agricultural environment, it is vital to comprehend these criteria.

2. Size and Growth Rate

The maximum sizes and rates of growth differ throughout species. The Whiteleg prawn is renowned for growing quickly; in only 4-6 months, it may reach market size. On the other hand, the Giant River Prawn grows more slowly but may develop to a large size, which might be useful in certain markets.

3. Feeding Patterns

Species also vary in their feeding habits. The Whiteleg prawn can adapt to several forms of diet, such as pellets and natural foods, and has a high feeding rate. However, since it is a more aggressive predator, the giant river prawn may need food that is newly produced or alive in order to develop to their full potential.

4. Resistance to Disease

The resistance to disease differs throughout prawn species. Despite its adaptability, the whiteleg prawn is prone to illnesses like Early Mortality Syndrome (EMS). Although the Black Tiger prawn is more resilient to illness in general, outbreaks must be carefully controlled. Planning efficient health management techniques is made easier by having a thorough understanding of each species' resistance to illness.

5. Lifecycle and Breeding

Requirements for breeding might also vary greatly. For the Giant River Prawn to successfully reproduce, several requirements must be met, such as the ideal salinity and temperature of the water. Although the Whiteleg prawn has a simpler breeding procedure, good larval survival rates still need to be ensured via careful management.

Selecting The Proper Species
1. Market Requirements

Understanding market demand is frequently the first step in selecting the appropriate species of prawn. Demand for valuable species like Black Tiger prawns might support more intricate growing practices. On the other hand, species with wider commercial appeal, like Whiteleg prawns, may be better suited for a more universal strategy.

2. surroundings

Evaluating your farm's environmental conditions is essential. For instance, the Giant River Prawn would be a good option if you have access to freshwater sources. The Whiteleg or Black Tiger prawns would be a better fit if your farm can keep consistent salinity and temperature levels.

3. Agricultural Resources

Think about the tools, technology, and knowledge that are accessible for farming. More sophisticated species, like the Kuruma prawn, could need more sophisticated methods and knowledgeable management. Conversely, the Whiteleg prawn is more adaptable and requires fewer complex settings to handle.

4. monetary considerations

The selection of species is significantly influenced by financial reasons. Due to their unique habitat requirements, certain species, like Black Tiger prawns, may demand a larger initial investment but might result in better revenues. With their speedier development rates and greater flexibility, species like the Whiteleg prawn may provide faster returns on investment.

5. Long-Term Objectives

Lastly, think about your long-term agricultural objectives. Investing in high-value species may be a good idea if your goal is to specialize and carve out a specialized market. Selecting a species as adaptable and scalable as the whiteleg prawn may be more beneficial if your goal is to create a farming operation that is both scalable and flexible.

CHAPTER THREE

Organizing Your Farm For Prawns

Choosing A Farm Location

The success of your prawn farm depends on where you decide to put it. A number of important requirements must be satisfied for the perfect location, such as accessibility, environmental factors, and closeness to essential resources.

environmental aspects to take into account

Assessing the possible places' environmental characteristics should come first. For prawns to flourish, the water temperature must fall between 25°C and 30°C. As such, choose a place where the weather often accommodates these temperatures. Take into account the site's vulnerability to severe weather conditions like flooding, wind, or heavy rain, since they might have a detrimental effect on the farm.

Availability of Water Sources

Having access to a dependable water supply is crucial to the upkeep of your prawn farm. Make sure the location has access to fresh, clean water or a dependable method of purifying the water. Being close to an established water supply system or a natural water body might be advantageous, but you should constantly look out for any hazards of pollution.

Infrastructure and Accessibility

Assess the site's suitability for building as well as continuing agriculture activities. For the purpose of transporting feed, equipment, and harvested prawns, the area needs to be conveniently accessible by road. A farm's operating effectiveness may be greatly impacted by the availability of infrastructure, such as roads, power, and communication services.

Compliance with Laws and Regulations

Do your homework on zoning laws and municipal rules pertaining to aquaculture before deciding on a place. Verify that the selected location satisfies all legal criteria and has the required permissions. Associations for aquaculture or local government agencies might provide advice on adhering to regulations.

Pond Design And Construction
Pond Layout and Design

The production and administration of your farm are significantly impacted by the design of your prawn ponds. Efficient water management, ideal prawn development, and simplicity of maintenance are guaranteed by proper design.

Size and Form of the Pond

Based on the size of your enterprise, choose the form and size of your ponds. Ponds may range in size from little, standalone features to massive, networked systems. While circular ponds might provide greater water circulation, rectangular or square ponds are often simpler to maintain. Make sure each pond has enough area for the prawns to develop and can be easily harvested.

Water Circulation and Flow

Maintaining water quality and making sure that nutrients and oxygen are distributed uniformly depend on efficient water flow and circulation. In order to promote water circulation and avoid stagnation, make sure your ponds have the right inlets and exits. Use aeration devices, such as paddles or diffusers, to raise the water's oxygen content and movement.

Selection of Soil and Liner

To stop water leaks and pollution, use the right soil and pond liners. Although clay soils are often used for their inherent sealing qualities, synthetic liners may also be utilized to provide further security. To ensure effective evacuation of water and collection of silt, make sure the bottom of the pond is sloped in the direction of the drain.

Building Supplies and Methods

When building a pond, choose sturdy materials like masonry, reinforced concrete, or premium liners. Use expert building methods to guarantee the pond structure's integrity. Maintain and examine ponds on a regular basis to catch problems early and save expensive repairs.

Management Of Water Quality

Keeping an eye on and preserving water quality.

Prawn growth and health depend on the maintenance of water quality. Maintaining ideal water quality and halting the spread of illness requires regular inspection and upkeep.

Important Water Quality Factors

Keep an eye on important water quality indicators including pH, temperature, dissolved oxygen, ammonia, nitrite, and nitrate concentrations. Make use of trustworthy testing tools and keep track of data on water quality. Regular assessments are essential for timely modifications since prawns need particular ranges for these criteria to grow.

Filtration and Water Treatment

Install water treatment systems to deal with problems like turbidity or excessive ammonia levels. Particulate matter and dangerous compounds may be eliminated from water with the use of filtration devices, such as mechanical filters and biofilters. To guarantee the efficacy of these systems, clean and maintain them on a regular basis.

Control and Prevention of Illnesses

Preventing disease outbreaks requires maintaining the quality of the water. Put biosecurity measures in place to reduce the possibility of disease introduction. Check prawns often for illness symptoms, and if needed, get advice from aquaculture experts on available treatments.

Recycling and Water Exchange

To maximize water utilization and minimize waste, create a strategy for water exchange and recycling. Replace some of the pond's water on a regular basis to replenish and dilute pollutants. Systems for recycling used water may help reduce environmental impact and increase resource efficiency.

Maintaining Records and Analysing Them

Maintain thorough records of all treatment procedures, water quality assessments, and pond management techniques. Examine these documents to see patterns and decide on water management tactics with knowledge. An atmosphere conducive to healthy and fruitful prawn farming would be ensured by constant observation and adjustment.

CHAPTER FOUR

Basics Of Prawn Hatchery

Equipment And Setup For Hatcheries

Establishing an ideal environment for the growth and development of prawns involves meticulous planning and evaluation of several aspects. Choosing the ideal site, planning the infrastructure for the hatchery, and making necessary equipment purchases are all part of the setup process.

Place and Design of the Facility

A prawn hatchery's location is essential to its success. It should be placed where there is little chance of pollution from outside sources and easy access to freshwater sources. Enough room should be available for the various phases of prawn growth, such as grow-out tanks, nurseries, and larval rearing.

The hatchery should be designed with distinct areas for the various phases of prawn growth. Appropriate filtration systems for maintaining water quality, temperature controls for controlling water temperature, and aeration systems for guaranteeing sufficient oxygen levels should be installed in each area.

Essential Tools

A prawn hatchery requires many kinds of equipment. Among them are:

1. Tanks and Ponds: The main tanks or ponds must be made big enough to hold prawns at different developmental stages. Larger volumes and more powerful filtration systems are needed for nursery and grow-out tanks, whereas fine mesh screens are necessary for larval tanks to stop contamination and escape.

2. filtering Systems: To rid the water of organic waste, sediment, and dangerous bacteria, high-quality filtering systems are required. These

systems aid in preserving the purity of the water and halting the spread of illness.

3. Aeration Devices: Aeration devices provide the water the essential oxygen, such as air pumps and diffusers. Prawns need sufficient aeration to be healthy and flourish.

4. Heating and Cooling Systems: The ideal development of prawns depends on temperature regulation. In colder areas, heating systems could be necessary, whereas in warmer ones, cooling systems might aid in preserving temperature stability.

. Tools for Monitoring Water Quality: It's critical to regularly check water quality factors including pH, salinity, and ammonia levels. Equipment like ammonia kits, pH meters, and salinity testers assist in making sure the water stays within acceptable bounds.

Upkeep and Sanitation

To combat disease outbreaks and maintain the prawns' health, the hatchery must be kept clean and hygienic. It is important to put in place regular cleaning schedules, equipment disinfection procedures, and appropriate waste management techniques.

Larval rearing and breeding

In prawn farming, breeding, and larval rearing are crucial phases that call for close attention to detail. Choosing broodstock, controlling breeding circumstances, and guiding larvae through their early phases of development are all part of these procedures.

The Selection And Management Of Broodstock

The quality of the broodstock has a major impact on breeding success. To breed, choose prawns that are disease-free, healthy, and possess the desired genetic qualities. It's important to acclimatize broodstock to the hatchery setting and offer them ideal circumstances for spawning.

Conditions for Breeding

Successful reproduction depends on spawning conditions being created. In the breeding tanks, keep the water at the proper temperature, salinity, and oxygen content. Spawning rates may also be increased by giving broodstock enough food and room.

The Larval Rearing Method

The fertilized eggs will develop into larvae after spawning. The larval stage is very delicate and needs certain circumstances in order to survive. To aid in the growth of larvae, keep an eye on and regulate the salinity, temperature, and quality of the water.

1. Use well-thought-out larval tanks with fine-mesh screens to keep insects from escaping.

Maintaining water quality requires appropriate filtration and aeration.

2. Feeding: A specific diet is needed for larvae, which usually consists of rotifers, microalgae, and other live feeds. To encourage healthy development, make sure a steady supply of high-quality feed is available.

3. Monitoring and Management: Keep an eye out for any anomalies or indications of illness in the larvae. In order to promote healthy growth, quickly implement any essential therapies and make any necessary environmental condition adjustments.

4. Making the Switch to Nursery: Larvae ultimately reach the juvenile stage as they continue to grow and develop. Move them carefully into nursery tanks, where they will develop further until they are prepared to be stocked in grow-out ponds.

Management Of Stocking And Nurseries

For prawns to grow and develop well, the stocking and nursery phases must be managed properly. During this stage, the nursery is set up, the prawns are stocked, and efficient growth management techniques are used.

Setting Up the Nursery

Make sure the ponds or nursery tanks are ready for the juveniles before adding any animals. Install filtration and aeration systems, evaluate water quality parameters, and clean and disinfect the tanks. Make sure there is enough room and supplies in the nursery for the prawns to develop.

Stocking Procedures

Care should be used while stocking to prevent stress and crowding. Juveniles should be added to the nursery tanks at the proper densities determined by the size and layout of the tanks.

The prawns should be gradually acclimated to their new surroundings in order to reduce shock and guarantee a seamless transfer.

Development Control

Check the prawns' development and well-being on a frequent basis. Make sure they eat a balanced diet and adjust feeding times according to their developmental stage. To keep water clean, keep an eye on things and change the settings as necessary. To maintain the best possible atmosphere for the nursery, do regular cleaning and maintenance.

Disease Control

Disease outbreaks may have a major effect on the survival and development of prawns. Adopt a proactive approach to disease control that includes routine health examinations, quarantine guidelines for newly acquired animals, and timely medical attention for any illnesses. Make sure that biosecurity protocols are implemented to stop the spread of illnesses.

Gathering and Changing

When the prawns are the right size, be ready to harvest them or move them to grow-out ponds. Utilize optimal techniques during harvesting to reduce damage and stress. Make sure the grow-out habitat is prepared to support their ongoing development and growth if they are transitioning.

CHAPTER FIVE

Nutrition And Feeding In Prawn Farming

Prawn Feed Types

For prawns to develop, be healthy, and be productive overall, they must be fed effectively. There are several varieties of shrimp feed available, each meeting the unique nutritional requirements and developmental phases of prawns.

Trade Feeds

Commercial shrimp feeds are made specifically to suit the nutritional needs of prawns during different phases of their development. They are available in several forms, including pellets, crumbles, and powders, to suit a variety of feeding habits and tank setups. Because they are easy to handle and store, pelleted feeds are widely utilized. To ensure that all prawns can reach the food, these pellets are designed to float or sink depending on how they are supposed to be used.

Organic Meals

Plankton, algae, and other naturally occurring aquatic creatures may be added to prawns' diets as supplements. These meals may resemble the prawns' normal eating habits and are high in vital nutrients. But unless properly handled, depending only on natural feeds may be difficult and could not result in a balanced diet.

Personalised Meals

Some farmers choose to mix additives such as fishmeal and soybean meal to make their own handmade prawn meals. This method may be economical and can be tailored to the individual requirements of the prawns. To guarantee the prawns have a balanced diet, nevertheless, one must have a solid grasp of nutritional needs and feed formulation.

Particularised Meals

Specialized meals are made with certain goals in mind, such as raising immunity, optimizing development rates, or increasing shell quality. Additional supplements, including vitamins, minerals, and growth boosters, are often included in these diets. Even though they may be more costly, they could have a big impact on the production and health of prawns.

. Feeding Methods And Schedules

To maximize prawn development and reduce wastage, an efficient feeding schedule, and appropriate feeding practices are essential.

Frequency of Feeding

The number of times a day that prawns are fed typically varies based on their size and age.

Compared to adults, larvae and juveniles may need to be fed more often. For smaller prawns, feeding may take place three to four times a day; for adult prawns, feeding might take place two to three times a day.

It's essential to modify the feeding schedule in accordance with the development stages and consumption rates of the prawns.

Feed Amount

To prevent overfeeding or underfeeding, it is essential to calculate the appropriate feed quantity. While underfeeding might impede development, overfeeding can cause problems with the quality of the water.

Offering a serving size of food that can be finished in 20 to 30 minutes is a smart idea. Maintaining ideal development requires keeping an eye on the prawns' reaction and modifying the amount of feed given in response.

Feeding Methods

A variety of feeding strategies may be used to guarantee effective feed utilization. By distributing feed uniformly across the surface of the tank or pond, broadcasting enables prawns to naturally graze. Although this technique is effective for pelleted diets, some feed may be wasted. Using automated feeders, which may deliver food at pre-set times to ensure constant feeding and cut down on labor, is another tactic.

Keeping an eye on meal consumption

It is crucial to regularly observe feed intake and prawn behavior in order to modify feeding procedures. It is possible to determine if the prawns are getting the proper quantity of feed and whether there are any indications of feed-related problems by keeping an eye on their feeding behavior. The feeding schedule and methods may be adjusted on time with the aid of this monitoring.

Dietary Needs

A balanced diet that promotes growth, reproduction, and general health must take into account the nutritional needs of prawns.

Macronutrients

Proteins, lipids, and carbs are among the macronutrients that prawns need for a balanced diet. Proteins are essential for the development of muscles and tissues; typical sources of protein include fishmeal and soybean meal. Carbohydrates are a secondary energy source, while fats provide energy and support many physiological processes. For optimum development rates and health, these macronutrients must be in the proper proportion.

tiny nutrients

Micronutrients, such as vitamins and minerals, are important for the growth and health of prawns. Vitamins A, D, and E are crucial for immune system performance, bone health, and eyesight. Magnesium, calcium, and phosphorus are among the minerals that are necessary for shell development and general metabolic functions. Making sure there are sufficient amounts of these micronutrients in the feed aids in the prevention of deficits and the enhancement of general health.

Vital Amino Acids

Proteins are made up of amino acids, and prawns need certain critical amino acids in particular to develop to their full potential. Arginine, methionine, and lysine are a few of them. Sufficient amounts of these amino acids in a balanced feed composition are necessary to promote normal growth and development.

Ratio of Feed Conversion and Digestibility

One of the most important factors in assessing the efficacy of feed is its digestibility. The optimal absorption and utilization of nutrients by prawns is ensured by high-quality meals with excellent digestibility. The effectiveness with which prawns convert feed into body weight is gauged by the feed conversion ratio or FCR. A reduced feed conversion ratio (FCR) signifies superior growth and feed efficiency.

Feeding and Water Quality

The prawn's capacity to process and use feed is directly impacted by the quality of the water. Ensuring optimum feed utilization and general prawn health requires maintaining appropriate water quality parameters, including pH, temperature, and dissolved oxygen levels. Healthy prawn development may be supported and feed-related problems can be avoided with routine monitoring and alterations to the water quality.

CHAPTER SIX

Nutrition And Feeding

Nutrition and feeding are essential components of a productive prawn farm. When these variables are properly managed, prawns develop, stay healthy, and produce as much as possible. This chapter examines several kinds of shrimp feed, feeding plans and methods, and the dietary needs crucial to prawn farming.

Prawn Feed Types

To choose the finest prawn feed choice for your farm, it is essential to understand the many varieties that are available. Prawn feed may be generically divided into a number of varieties, each with unique qualities and advantages.

1. Commercial Feed: Feeds designed for commercial purposes are tailored to fulfill the nutritional requirements of prawns. These feeds are available in pellet, crumble, and powder form. They are often enhanced with vital

vitamins, minerals, and nutrients. Prawns are guaranteed a healthy diet with commercial meals since they are convenient and reliable.

2. native Feed: In the prawns' native environment, natural feeds include things like algae, plankton, and tiny crustaceans. These meals support natural foraging behaviors and provide a variety of nutrients. Natural foods provide advantages, although there may be variations in their quality and availability.

3. Homemade Feed: To make homemade feeds, a variety of components are combined, including vegetable matter, fish meal, and soy protein. This kind of feed is customizable based on individual dietary needs and might be a reasonably priced choice. To guarantee nutritional balance, nevertheless, meticulous formulation and oversight are necessary.

4. Supplementary Feed: To improve the nutritional profile, supplemental feeds are

employed in addition to basic feed sources. Additives such as vitamins, minerals, or growth boosters could be among them. Supplementary meals may improve general prawn health or aid with particular deficits.

5. Feed: Depending on its buoyancy, the feed may be categorized as floating or sinking. Because floating feeds float to the top, prawns can more easily eat them, and farmers can keep an eye on their eating habits. Prawns that eat closer to the substrate may benefit from sinking meals, which, on the other hand, settle near the bottom.

Feeding Methods And Schedules

Optimizing prawn development and reducing waste requires using proper procedures and putting in place an efficient feeding plan.

1. Feeding Frequency: To make sure prawns have enough nutrients, they should be fed many times a day. Depending on the prawns'

size and age, feeding usually takes place two to three times a day. Larger prawns can be fed less often, while younger prawns may need to be fed more frequently.

2. Feed Quantity: The size, age, and farm environment should all be taken into consideration when determining how much feed to give prawns. While underfeeding might impede development, overfeeding can cause waste and problems with water quality. Generally speaking, feed should be provided so the prawns may finish in a few hours.

3. Feeding Methods: To maximize feed efficiency, a variety of methods may be used. By distributing feed uniformly over the water's surface, broadcasting induces feeding in all prawns. Spot feeding, which is helpful in bigger tanks or ponds, focuses on certain locations. The feed may also be precisely delivered by automated feeders at predetermined intervals.

4. Feed Intake Monitoring: To make necessary adjustments to feeding techniques, it is essential to periodically observe prawn behavior and feed intake. It is possible to determine if adjustments are required by keeping an eye on the speed at which the prawns devour the feed and any changes in their eating habits.

5. Changing Feed Types: Depending on the prawn's stage of development, the kind of feed may need to be changed. For example, bigger prawns can eat coarser pellets, whereas tiny prawns would need finely powdered feed. For optimal consumption, the feed size must correspond to the size of the prawn's mouth.

Dietary Needs

Fulfilling the dietary requirements of prawns is essential for their development, well-being, and general functionality. A meal rich in proteins, fats, carbs, vitamins, and minerals is necessary for prawns.

1. Proteins: Because they are the building blocks of tissues and enzymes, proteins are necessary for growth and development. High-quality protein sources, such as fish or soybean meal, are necessary for prawns. The prawn's development stage and the surrounding circumstances should be taken into consideration while adjusting the feed's protein content.

2. Lipids: Also referred to as fats, lipids provide a concentrated form of energy and are essential for many metabolic functions. They also contribute to the prawn's exoskeleton's development. Maintaining the proper amount of fat in the diet supports both general health and energy requirements.

3. Carbohydrates: Prawns primarily get their energy from carbohydrates. They are nonetheless necessary for sustaining energy levels and general health, even if they are not as vital as proteins and fats. It is important to

balance the feed's carbohydrate content to prevent excess waste and possible problems with water quality.

4. Minerals and vitamins: Although needed in lower amounts, vitamins and minerals are essential for a number of physiological processes, such as immune response and bone formation. Vitamins A, D, and E, as well as minerals like calcium and phosphorus, are often added substances.

5. Fibre and Other Additives: Fibre facilitates digestion and supports a wholesome environment in the gut. Probiotics and growth boosters are examples of additional additives that may be used to assist general health and development.

Prawn farmers may improve the health and production of their stock by using efficient feeding schedules and practices, choosing high-quality feed, and making sure that all nutritional needs are satisfied. The secret to effective prawn farming is regular monitoring and modifications depending on the demands of the prawns and the farm environment.

CHAPTER SEVEN

Water Resources Management

Parameters Of Water Quality

Since water quality directly affects the health and development of prawns, it is essential for effective shrimp farming. To guarantee a profitable aquaculture environment, it is important to comprehend and regulate the parameters of water quality.

1. The temperature

An essential factor in prawn farming is temperature. Warm water is ideal for most prawns, with ideal temperatures falling between 28°C and 32°C. When this range is departed from, the prawns may get stressed, which might slow their development and make them more prone to illness. It's crucial to regularly check the water's temperature using dependable thermometers. Installing heaters or coolers may help maintain stability when the temperature veers outside the ideal range.

2. pH Scales

The prawns' general health and nutritional availability are impacted by the pH of the water. A pH range of 7.5 to 8.5 is commonly preferred by prawns for their habitat, which is somewhat alkaline. It is essential to do routine pH testing using pH meters or test kits. Balance may be restored if the pH deviates from the intended range and is adjusted using pH buffers or natural therapies like lime.

3. Oxygen That Has Disintegrated

The breathing and general health of prawns depend on dissolved oxygen (DO). Prawns need dissolved oxygen at high concentrations—usually more than 5 mg/L. High death rates and poor growth may result from low DO levels. Diffusers and air pumps are examples of aeration devices that may raise the water's oxygen content. Maintaining ideal oxygen levels is ensured by routinely testing DO levels using DO meters.

4. Levels Of Ammonia And Nitrite

Waste and uneaten feed may cause harmful substances like ammonia and nitrite to build up in the water. It is recommended to maintain ammonia levels below 0.1 mg/L and nitrite levels below 0.2 mg/L. It is essential to use water testing kits to keep an eye on these characteristics. Regular water changes and the installation of effective filtration systems may aid in the management and reduction of these hazardous substances.

5. Acidity

Prawns' ability to osmoregulate is impacted by salinity. While salinity needs vary throughout prawn species, most do well in salinity ranges of 10 to 20 ppt (parts per thousand). Salinity meters or refractometers should be used to monitor salinity. To maintain the appropriate salinity levels, saline water may be added or diluted.

Systems For Filtration And Aeration
1. Systems of Filtration

By eliminating suspended particulates and biological trash, filtration systems are crucial for preserving the purity of water. In prawn farming, a variety of filtering techniques are used, including:

• Mechanical Filters: These filters rid the water of physical particles. Sand filters and cartridge filters are examples of common kinds. Particulate matter is reduced and pure water is maintained with the use of mechanical filtering.

• Biological Filters: These filters use good microorganisms to degrade toxic substances like nitrites and ammonia. In order to support the natural breakdown of waste materials in the water, biofilters are essential for creating a healthy microbial ecology.

• Chemical Filters: These filters, which include activated carbon filters that aid in eliminating organic chemicals and odors, employ chemical media to remove dissolved contaminants. For the best possible water quality, chemical filtration may be used in addition to mechanical and biological filtering.

2. Systems of Aeration

To guarantee that the water has the right amount of oxygen, aeration devices are essential. Water circulation is improved and oxygen diffusion is enhanced by proper aeration. Typical aeration systems consist of:

• Air Pumps: Air pumps employ diffusers or air stones to add air to the water. They work well to raise oxygen concentrations and stimulate surface agitation.

- Diffusers: By introducing tiny bubbles into the water, air diffusers increase the surface area available for the exchange of oxygen. When regular aeration is required in big ponds or tanks, they are very helpful.

- Surface Aerators: These gadgets stir up the water's surface to help it absorb oxygen from the surrounding air. Larger systems with a high oxygen demand often use surface aerators.

3. Upkeep of Systems

The effectiveness of filtration and aeration systems depends on their maintenance. Maintaining maximum performance of filters is ensured by regular cleaning and examination. Examine the filter material for damage or obstructions, and replace it as necessary. In order to avoid debris accumulation, air stones or diffusers should be cleaned on a regular basis and aeration systems should be inspected to ensure optimal performance.

Frequent Inspection And Upkeep

1. Standard Inspections

It is important to conduct routine monitoring of water quality indicators in order to identify any abnormalities that may have an influence on the health of prawns. Set up a regular testing plan for salinity, dissolved oxygen, pH, ammonia, and nitrite in water. To monitor changes over time, use dependable testing tools and document the findings.

2. Upkeep of the Equipment

For filtration and aeration systems to remain functional, proper maintenance is required. As advised by the manufacturer, change the media and clean the filters. Examine aeration systems for wear and tear or obstructions. To avoid compromising the quality of the water, routinely check all equipment for indications of failure and take quick corrective action when necessary.

3. Changes in Water

Water quality is maintained and accumulated waste items are diluted with regular water changes. Based on the size of the system and the prawn stocking density, determine the proper frequency and amount of water changes. To prevent significant changes in water quality metrics, do partial water changes.

4. Health Surveillance

Regularly check the prawns' health for any indications of illness or stress. Prawns in excellent health are a sign of high-quality water. Take remedial action and look into any problems with the water quality if any anomalies are found. Make sure that no changes or treatments applied to the water may adversely affect the aquatic environment's overall equilibrium.

5. Maintaining Records

Keep thorough records of all system modifications, equipment upkeep, and water quality data. Maintaining reliable records is beneficial for seeing patterns, evaluating the success of management techniques, and formulating well-informed plans for future advancements.

Aeration, filtration, and water quality systems need to be regularly maintained if prawn farming is to be effective. Prawns may flourish in a steady, healthy habitat if you follow these guidelines.

CHAPTER EIGHT

Water Resources Management

Overview of Water Management in Prawn Production

The key to a profitable prawn farming operation is effective water management. In addition to providing prawns with ideal growing circumstances, proper management guards against illnesses that might result from low water quality. Numerous crucial elements, like as frequent monitoring and maintenance, filtration and aeration systems, and water quality metrics, are involved in understanding how to maintain and manage water in your prawn farm.

Parameters of Water Quality

Prawns need clean water to thrive and stay healthy. To keep an aquatic environment healthy, a number of criteria must be regularly observed. These consist of ammonia levels, temperature, salinity, pH, and dissolved oxygen.

• Temperature: Certain temperature ranges are ideal for prawn growth. Temperature extremes may stress prawns and stunt their development. Prawns typically want their temperature to be between 25°C and 30°C. It takes consistent observation with a trustworthy thermometer to maintain the water in this ideal range.

• Salinity: Depending on the kind of prawn being farmed, salinity levels should be managed. For example, the salinities that marine prawns need are typically between 15 and 30 ppt (parts per thousand). Variations outside this range may have an effect on the development and health of prawns. Salinity levels are measured and adjusted using a refractometer or salinity meter.

• pH Levels: Prawn metabolism and nutrient availability are impacted by the pH of the water. Most prawn species like pH values between 7.5

and 8.5. Maintaining these levels and making necessary adjustments with acidic or alkaline solutions is made easier with routine pH testing.

- Dissolved Oxygen (DO): Prawns need sufficient oxygen to breathe and develop. It is best to maintain oxygen levels beyond 5 mg/L. Systems for aeration aid in maintaining high dissolved oxygen concentrations. Frequent DO testing guarantees that oxygen concentrations stay within the necessary range.

- Ammonia Levels: High amounts of ammonia, a result of uneaten feed and prawn waste, may be harmful. It is best to keep ammonia levels as low as possible, preferably less than 0.5 mg/L. Ammonia levels may be managed with the use of biological filters and regular water changes.

Systems For Filtration And Aeration

By eliminating pollutants and guaranteeing sufficient oxygen levels, filtration and aeration systems are essential to preserving the quality of water.

- Filtration Systems: To remove organic matter and solid waste from water, filtration is necessary. It is possible to utilize a variety of filters, such as chemical, biological, and mechanical filters. Physical particles are removed using mechanical filters, such as sand filters and sieves. Beneficial microorganisms are used by biological filters to convert hazardous ammonia into less damaging compounds. Certain impurities may be eliminated by chemical filters by adsorption or chemical reactions.

- Aeration Systems: Aeration systems, which include diffusers, air pumps, and aerators, aid in raising the water's dissolved oxygen content. Diffusers are pushed by air pumps, which produce bubbles that improve the exchange of oxygen. Aerators are a useful tool for increasing the oxygenation and movement of water, particularly in big tanks or ponds. Maintaining aeration equipment on a regular basis guarantees effective functioning and guards against oxygen shortages.

Frequent Inspection And Upkeep

Maintaining water quality and a safe environment for prawn farming needs regular monitoring and maintenance.

• Monitoring Water Quality: To identify and quickly resolve any problems, regular monitoring of water parameters is necessary. Maintaining ideal conditions may be aided by scheduling regular testing of the temperature, salinity, pH, dissolved oxygen, and ammonia levels. Finding patterns and making the required corrections may be made easier by using trustworthy testing apparatus and maintaining results documentation.

• Maintaining Filtration Systems: To maintain the effectiveness of filtration systems, regular cleaning and service are required. While biological filters may need to have their filter material replaced on a regular basis, mechanical filters should be cleaned to get rid of collected dirt. It is necessary to examine chemical filters for saturation and to regenerate or replace them as necessary.

• Maintenance of Aeration Systems: To make sure they are operating properly, aeration systems should be examined on a regular basis. It's critical to look for any obstructions or issues with the air pumps and diffusers. The aeration system's efficiency will be maintained, and oxygen shortage will be avoided, by cleaning or replacing components as required.

• Water Changes: Keeping the water at its best and minimizing waste buildup are two benefits of regular water changes. The size of the farm, the quantity of prawns, and the water quality all affect how often and how much water is used. Maintaining a healthy atmosphere may be aided by partial water changes in conjunction with appropriate filtration.

To sum up, proficient water management in prawn farming entails comprehending and regulating water quality criteria, putting in place appropriate filtration and aeration systems, and carrying out routine maintenance and monitoring. Following these guidelines will help you raise prawns in a stable, healthy environment that will provide large harvests and profitable farming.

CHAPTER NINE

Water Resources Management

The efficient management of water is essential for prawn farming. Prawns develop, survive, and are generally of higher quality when they are kept in a healthy habitat, which is ensured by proper management. The fundamentals of water management are covered in detail in this chapter, including filtration and aeration systems, regular maintenance, and water quality measures.

Parameters Of Water Quality

A vital component of prawn growth and health is water quality. It includes a number of important factors, all of which are vital to preserving an environment that is favorable for prawns.

1. Temperature Control: One of the most important factors in prawn farming is temperature. As ectothermic creatures, prawns' body temperature is influenced by their

surroundings. Most prawn species like temperatures between 24°C and 30°C. Straying from this range may cause stress to the prawns, which might impact their health and development rates. Maintaining the ideal range may be aided by routinely checking and modifying the water's temperature with the use of heaters or chillers.

2. pH Levels: The prawns' general health and the availability of nutrients are impacted by the pH level of the water. In general, slightly alkaline water with a pH range of 7.5 to 8.5 is ideal for prawn growth. Conditions that are too acidic or too alkaline might hinder development and make people more prone to illness. To maintain the optimal range, it's critical to measure pH levels often and employ pH adjusters as needed.

3. Salinity Prawn osmoregulation and general health are impacted by the salinity levels in the water. Depending on the kind of prawn, there

are several ideal salinity ranges, but generally speaking, they are between 5 and 30 ppt (parts per thousand). It is crucial to conduct routine assessments of salinity and make necessary modifications to guarantee that prawns stay within their ideal salinity range.

4. Dissolved Oxygen (DO): Prawns' ability to breathe depends on dissolved oxygen. Low oxygen levels may cause stunted development and a high death rate. It is advised to keep DO levels above 5 mg/L. Appropriate aeration techniques—which will be covered in more detail—can help accomplish this.

5. Levels of Ammonia, Nitrite, and Nitrate Nitrate, although less toxic, may still have an impact on shrimp health when present in large amounts. Ammonia and nitrite are hazardous to prawns even at low concentrations. It is important to do routine testing and maintenance on these parameters. With the use of biological filtering systems, hazardous ammonia and

nitrite may be changed into less dangerous nitrate, which can then be eliminated by water exchanges or other procedures.

6. Suspended Solids and Turbidity: Turbidity is the cloudiness of water brought on by suspended solids. Excessive turbidity may hinder light penetration and interfere with aquatic plants' capacity to photosynthesize, which in turn affects the availability of nutrients and oxygen. Clear and healthful water may be maintained by routinely checking and regulating turbidity using filtration devices and frequent water changes.

Systems For Filtration And Aeration

Water management in prawn farming requires effective filtration and aeration systems. They promote the general health of prawns, guarantee appropriate oxygen levels, and preserve the purity of the water.

1. Filtration Systems: Filtration systems eliminate dangerous chemicals, organic materials, and suspended particles from water. A variety of filters may be applied, such as:

• Mechanical Filters: Particles in the water are physically removed by these filters. They consist of drum filters and mesh screens, which hold onto solids and keep them from building up in the pond.

• Biological Filters: These filters use good microorganisms to degrade toxic chemicals like nitrite and ammonia. Biofilters and moving bed bioreactors are two examples.

• Chemical Filters: These filters eliminate pollutants via chemical processes. One popular kind of filter that absorbs contaminants and organic chemicals is activated carbon.

2. Aeration Systems: Aeration systems are necessary to keep the water's dissolved oxygen

levels at a sufficient level. There are many varieties of them, such as:

• Air Stones and Diffusers: These appliances increase oxygen circulation and transmission by releasing tiny bubbles into the water. They are often used as additional aeration or in smaller systems.

• Surface Aerators: By stirring up the water's surface, these gadgets improve oxygen exchange with the atmosphere. They work well with high-density stockings and bigger ponds.

• Bottom Aerators: These devices provide uniform oxygen distribution throughout the pond by mixing the water column from the bottom. In deeper or bigger ponds, they are helpful.

3. Automation and Integration Combining automated controls with filtration and aeration systems may boost productivity and cut down on human observation. In order to maintain ideal water quality, automated systems may

regulate filtering cycles and alter aeration rates depending on current oxygen levels.

Frequent Inspection And Upkeep

For prawn farming to be successful over the long term, filtration, aeration, and water quality factors must be regularly monitored and maintained. The finest methods for efficient management are described in this section.

1. Regular water parameter testing is essential for spotting problems early on and fixing them before they have an adverse effect on prawn health. Temperature, pH, salinity, dissolved oxygen, ammonia, nitrite, nitrate, and turbidity should all be tested often. Track changes over time by using dependable test kits and documenting the findings.

2. System Inspections: To make sure filtration and aeration systems are operating correctly, do routine inspections. Examine for wear and tear, blockages, and any indications of a

malfunction. To keep components operating at peak efficiency, clean or replace them as necessary.

3. Water Exchange and Treatment: Water quality is maintained and stored contaminants are diluted by periodic water exchanges. Based on the pond's size and stocking density, determine the appropriate water exchange frequency and volume. Additionally, if required, treat the water with conditioners or dechlorination agents.

4. Schedule of Maintenance: All systems and equipment should have a schedule of maintenance established. Incorporate duties like sensor calibration, aeration device inspection, and filter cleaning. Frequent maintenance guarantees that the systems function well and help to avoid malfunctions.

5. Keeping Records: Keep thorough records of all system maintenance, modifications, and

water quality data. For the purpose of identifying patterns, addressing problems, and deciding on appropriate water management strategies, this data is invaluable.

This in-depth discussion of water quality criteria, filtration, and aeration systems, and the need for routine maintenance and monitoring should provide readers with a thorough grasp of water management in prawn farming.

CHAPTER TEN

Legal And Environmental Aspects

Sustainability And Its Effect On The Environment

Shrimp farming, sometimes referred to as "prawn farming," is a profitable and rapidly growing sector of the economy that has a big impact on the environment and other ecosystems. Responsible aquaculture depends on adopting sustainable techniques and being aware of the environmental effects of the growing worldwide demand for prawns.

Ecosystem Disturbances

Prawn farms have the potential to disturb ecosystems in a number of ways. The change in natural environments is one of the main problems. In order to provide room for prawn ponds, mangroves, marshes, and coastal regions are often destroyed. Deforestation impacts these ecosystems' natural filtering systems in addition to lowering biodiversity. For example, mangroves are essential for maintaining a variety of marine life and preventing erosion along coasts.

Management of Water Quality

A key element of sustainable prawn farming is water quality. When farmed prawns are introduced into aquatic settings, the chemistry of the water may be affected, which can result in problems including excess nutrients and higher quantities of organic waste. Adequate management techniques are necessary to avoid the build-up of toxic materials such as ammonia and nitrites, which may have detrimental effects on both domesticated and untamed animals. To maintain a balanced and healthy environment, water has to be regularly monitored and treated.

Management of Effluent

The waste product from prawn farming activities is called effluent or wastewater. This effluent may include high concentrations of nutrients and pathogens if it is not adequately treated, which might contaminate nearby water bodies. Using effluent management techniques, such as settling ponds and filtering systems, to treat and recycle wastewater is a component of sustainable prawn farming operations. By doing this, the effect on the environment is lessened and water quality is preserved.

Disease Control

In prawn farming, disease outbreaks are frequent and may result in large-scale damage to the environment and the economy. Using chemicals and antibiotics to treat illnesses may result in resistance and other environmental problems if not managed appropriately. To reduce the negative effects on the environment, integrated disease management techniques are crucial. These include using disease-resistant prawn strains, maintaining good farm cleanliness, and implementing biosecurity protocols.

Eco-Friendly Meal Techniques

There are environmental effects associated with the feed used in prawn farming. Fishmeal and fish oil are common ingredients in commercial prawn diets, and they may lead to overfishing and the loss of marine resources. Sustainable feed techniques include creating substitute feeds that lessen dependency on wild-caught fish or obtaining feeds from ethical fisheries. This promotes the long-term sustainability of prawn farming in addition to aiding in the conservation of marine resources.

Regulations Needed

In order to guarantee that aquaculture activities do not negatively impact the environment or public health, regulations pertaining to prawn farming have been put in place. Although these laws differ from nation to nation and area to region, they often address issues like food safety, animal welfare, and environmental preservation.

Rules Regarding the Environment

The goal of environmental laws is to reduce the ecological footprint of prawn farming. These might include limitations on the size and location of farms, the need for environmental impact studies prior to farm formation, and rules pertaining to waste management and effluent treatment. Following these rules contributes to the promotion of sustainable agricultural methods and the prevention of environmental harm.

Standards for Animal Welfare

The purpose of animal welfare guidelines is to guarantee that prawns raised in farms are treated humanely. These guidelines include things like handling procedures, antibiotic and chemical use, and livestock densities.

Respecting animal welfare guidelines encourages moral farming methods and enhances the general well-being and yield of the prawns.

Standards For Food Safety And Quality

Prawns must meet strict requirements for food safety and quality in order to be considered safe for ingestion. These guidelines include the use of pesticides and medicines, as well as the need to keep an eye out for and manage any pollutants. Market criteria for prawn products are met and foodborne diseases may be avoided by adhering to food safety rules.

Licenses and Permits

Permits and licenses from regional or federal agencies are usually needed for prawn farming enterprises. Permissions for building farms, using water, and discharging wastewater are a few examples of these permissions. To operate legally and prevent future fines or penalties, it is essential to get and maintain the required permissions.

Observation and Documentation

Prawn growers are often required by regulations to keep a close eye on and report on a variety of operational elements. Data on water quality, effluent discharge levels, and prawn farm health records are a few examples of this. Frequent reporting guarantees openness and enables regulatory bodies to monitor adherence to safety and environmental regulations.

The Best Methods for Adherence

Prawn producers should use best practices that support sustainability and regulatory standards in order to guarantee adherence to legal and environmental obligations.

Using Sustainable Agriculture Methods

The environmental effect of prawn farming may be reduced by putting sustainable farming practices into practice, such as integrating pond and mangrove management systems. With the help of mangroves and other plants, these strategies may be used to create a balanced environment where prawns can benefit from natural filtration and protection.

Frequent Instruction and Training

Maintaining compliance with standards and best practices requires frequent training and instruction for farm personnel. Training courses on environmental management, animal welfare, and food safety should be included to make sure that employees are competent and able to follow rules.

Making Use of Innovation and Technology

Innovation and technological advancements may significantly improve compliance efforts. For instance, sophisticated feed formulations may lessen the environmental effect, while automated monitoring systems can measure water quality factors in real-time. Investing in technology increases agricultural productivity overall and aids in compliance maintenance.

Regularly Carrying Out Audits

To find opportunities for improvement and make sure that standards are constantly followed, regular audits of agricultural operations and regulatory compliance are essential. Internally or by other organizations, audits provide an unbiased evaluation of the farm's compliance with regulatory and environmental standards.

Assisting Stakeholders

Maintaining openness and fostering trust requires active engagement with stakeholders, such as local communities, environmental organizations, and regulatory bodies. The prawn farming sector as well as the larger ecosystem may both benefit from collaborative efforts that result in improved practices and solutions that meet environmental and regulatory problems.

CHAPTER ELEVEN

Increasing The Size Of Your Prawn Farm

To achieve long-term success, scaling and growing your prawn farm requires careful strategy, execution, and investment. This thorough guide will assist you in navigating this procedure.

Planning And Growth Strategies
Evaluating Ongoing Operations

Assess your present agricultural activities before scaling. Examine your ability to produce, the way you manage water quality,

and the steps you take to prevent illness. Determine what needs to be improved upon and provide a starting point for future development. You may use this evaluation to ascertain the resources needed and if expanding is feasible.

Establishing Expansion Objectives

Establish quantifiable, precise objectives for your development. Do you want to extend into new geographical areas, diversify your species, or enhance the amount of products you produce? Establish clear objectives for each aim, such as adding two new sites over the next three years or boosting prawn output by 50%.

Formulating a Strategic Plan

Make a thorough strategic plan that outlines the steps you will take to meet your development objectives. Timelines, resource allocation, and market research should all be part of this strategy. Think about things like manpower requirements, equipment demands, and the location of additional ponds. Make sure your strategy takes into account any obstacles that may arise and has backup plans.

Putting Money Into Infrastructure

Investing heavily in infrastructure is necessary for scaling up. Determine if more ponds, filtration units, and feed storage facilities are required. Make plans to upgrade current systems and install new equipment in order to accommodate higher output. Invest in technologies like automated feeding systems and water quality monitors that increase productivity and save labor expenses.

Education and Training

Make sure your staff has the training and experience necessary to oversee more complex operations as you grow. Continually educate employees on subjects including disease control, biosecurity, and operating procedures. Invest in the leadership development of key personnel so they can successfully manage additional teams and facilities.

Observation and Assessment

Establish a reliable tracking system to keep tabs on the development of your growth objectives. Review performance indicators on a regular basis, including survival rates, feed conversion ratios, and growth rates. Utilize this information to modify your tactics and quickly resolve any problems. Frequent assessments will assist you in staying on course and making wise choices.

Financial Management And Investment
Comprehending Capital Needs

Growing your prawn farm is going to cost a lot of money. Determine the amount of capital required for operating expenses, equipment, and infrastructure. Take into account both upfront expenditures and recurring costs, such as labor and maintenance. Create a thorough budget that takes into consideration all possible expenses.

Acquiring Financial Support

Look into your financing options to support your growth.

Personal savings, grants, loans, and venture capitalist investments are a few examples of these. Create a thorough company strategy and financial estimates to show prospective investors or lenders. Emphasize the development potential, track record, and Hazard Assessment of your farm.

Tactics.

Practices of Financial Management

To continue expansion, effective financial management is essential. Put accounting processes in place to monitor revenue, costs, and earnings. To make sure you remain within your budget and hit your financial goals, evaluate your financial accounts and performance indicators on a regular basis. To evaluate the state of your company and make wise choices, use financial ratios.

Measures for Cost Control

Find areas where costs might be reduced without compromising quality.

Ask vendors to reduce their prices on feed and equipment. Reduce utility costs by putting energy-efficient techniques into effect. To increase profitability, evaluate and tweak your cost-control strategies on a regular basis.

Hazard Assessment

Create a risk management strategy to deal with any monetary hazards. This plan needs to include liability, equipment, and property insurance. To prepare for unforeseen costs or economic downturns, think about establishing an emergency fund. Regularly review and adapt your risk management strategies to reflect changing conditions.

..Investigating Novel Markets And Prospects

Analysis and Research on the Market

For your prawn farm, do in-depth market research to find new chances. Examine market pricing, customer preferences, and demand patterns. Investigate prospective markets, including regional, global, and local choices. Recognize the market dynamics and regulatory requirements in each possible market.

Increasing the Variety of Products Offered

To appeal to a wider spectrum of consumers, think about expanding the variety of products you sell. Examine value-added goods including prepared dinners, specialty items, and processed prawn products. Diversification may improve income sources and reduce market volatility-related risks.

Creating Alliances

Form strategic alliances with retailers, distributors, and other interested parties to increase your market penetration. Work together with groups and associations in your sector to raise your profile and reputation. You may expand your company's chances and get access to new markets by cultivating excellent partnerships.

Examining Possibilities for Export

Assess the possibility of supplying your prawns to foreign markets. Examine trade agreements, export laws, and consumer demand in the nations you are targeting. Create a plan for expanding into new markets that takes local regulations into account and establishes distribution routes.

Creative Marketing Techniques

Use creative marketing techniques to advertise your prawn farm and its offerings. To reach more people, make use of internet platforms, social media, and digital marketing. Engage in industry gatherings and trade exhibits to present your goods and make connections with possible customers.

Conclusion

Considering the success of Prawn Farming

As our investigation into prawn farming for novices draws to a close, it's important to take stock of our shared experience. Prawn farming is a fulfilling endeavor that may become prosperous and successful with proper planning and effort. The abilities and understanding you have gained from this course are just the start of your prawn farming career. By grasping the principles, you have established a solid basis for your future success, from feeding and disease control to pond building and location selection.

Highlighting Important Lessons

We have discussed several important facets of prawn farming in this book. The significance of selecting the appropriate species and comprehending their particular needs, the need to preserve ideal water quality, and the need to put into practice efficient feeding and management techniques are among the most important lessons learned. Every one of these components is essential to maintaining the well-being and efficiency of your prawns. You may raise your chances of having a profitable crop and a successful company by following these guidelines.

Ongoing Education and Adjustment

The prawn farming sector is a dynamic one, with ongoing developments in techniques and technology. Maintaining current knowledge of new advancements and adjusting to modifications are essential for sustained success. To keep current on developing trends and best practices, think about networking with seasoned farmers, attending seminars, and joining industry associations. You may enhance your agricultural methods and deal with any obstacles by adopting an attitude of constant learning and adaptation.

Overcoming Obstacles

Prawn farming is no different from other agricultural endeavors in that it presents some problems. Problems like epidemics, variations in the quality of the water supply, and shifts in the market may be very difficult. But now that you have the newfound information and tactics at your disposal, you are more prepared to face these difficulties. You may reduce risks and guarantee the sustainability of your prawn farming operations by using proactive management, frequent monitoring, and professional assistance.

Defining objectives and assessing progress

Establishing measurable objectives for your prawn farming endeavor is crucial to tracking your progress and maintaining motivation. Having a roadmap will help you stay focused and give you a feeling of success, regardless of whether your objectives are monetary targets, production quantities, or personal fulfillment. You may remain on course and accomplish your goals by routinely evaluating your progress and making necessary adjustments to your methods.

Creating a Network and Looking for Help

Creating a network of suppliers, other prawn farmers, and industry professionals may be a great way to get advice and ideas. Through networking, one may exchange expertise, find answers to shared issues, and open doors to new possibilities. Never be afraid to ask for help and work together with other professionals in the field. Developing solid connections with other members of the prawn farming community may increase your chances of success and provide a safety net when things go tough.

Last Word of Encouragement

Recall that success in prawn farming takes time as you set out on your quest. It takes commitment, endurance, and a willingness to adapt and develop.

Accept the difficulties as chances for personal development, and acknowledge your accomplishments along the road. You have a strong base to work from thanks to the knowledge and ideas in this book. Have faith in your skills, stick to your objectives, and enjoy the fulfilling experience of raising prawns.

THE END

www.ingramcontent.com/pod-product-compliance
Lightning Source LLC
Chambersburg PA
CBHW071404220526
45469CB00004B/1157